NEDDY THE FORG

EVERYDAY SCIENCE STORIES

Health

BY THE SAME AUTHOR
For Children aged 3 to 7

Neddy the Forgetful Teddy Series: Everyday Science Stories

- *Air and Water* - *Animals and Plants* - *Forces*
- *Light and Sound* - *Health*

Ten Minute Tales

For Children aged 9 to 13 year old

The Sandmouse

The Tale of Thomas Thixendale
with Ian Durrant

Alien Invasion (short stories)

Young Adults

The Aurora Children

NEDDY THE FORGETFUL TEDDY
EVERYDAY SCIENCE STORIES
Health

by
Mavis Brown

Published by Mavis Brown 2024

All rights reserved. This book or any portion thereof may not be reproduced or used in any manner whatsoever without the express written permission of the publisher except for the use of brief quotations in a book review or scholarly journal.

Text Copyright © 2024 Mavis Brown
Cover and Illustrations © 2024 Mavis Brown

ISBN 978-1-326-89607-2

The right of Mavis Brown to be identified as author of this work has been asserted in accordance with Section 77 of the Copyright, Designs and Patents Act 1988.

First Printing: 2024

Published by Mavis Brown, Our House, The Green, Fritchley, Belper. Derbyshire DE56 2FW. UK

❈ ❈

INTRODUCTION

Neddy the Forgetful Teddy Series: Everyday Science Stories will be an invaluable resource for parents, teachers and carers of 3 to 7 year old children. They can cover the UK Foundation Stage Curriculum, for 3 to 5 year old children, and Key Stage 1 Curriculum in Science for 5 to 7 year old children.

The author, **Mavis Brown**, has taught Science in English schools to children of all abilities and ages for more than thirty years.

Neddy Teddy is a kind, elderly, toy bear who is rather forgetful and this causes him problems. His exploits with his friends show that science concepts can be found in everyday life.

This book about **Health** contains the following stories:

- **Neddy has trouble with his teeth** is about caring for your teeth and visiting the dentist.

- **Neddy takes Eli to the optician** is about going to the optician to check your eyesight.
- **Neddy has the flu** is about looking after yourself when you get ill.
- **Milly and Billy need a vaccination** is about having inoculations to prevent getting viral illnesses.
- **Neddy helps Percy to get fit** is about how your heart beats when you exercise.

Activities linking with these stories can be found in the following books, written by the author for Brilliant Publications, which cover the UK National Curriculum for Early Years (2012):

- *Physical Development with Expressive Arts and Design,*
- *Personal, Social and Emotional Development with Understanding the World and Mathematics.*

CONTENTS

	Page
Introduction	v
Neddy has trouble with his teeth	9
Neddy takes Eli to the optician	17
Neddy has the flu	23
Milly and Billy need a vaccination	29
Neddy helps Percy to get fit	35

NEDDY HAS TROUBLE WITH HIS TEETH

Neddy Teddy looked out of his window. The sun was rising and the birds were singing loudly.

Neddy had not slept a wink of sleep all night. He had toothache.

"What can I do about my teeth?" he asked himself. "I wonder if any of my friends have trouble with their teeth?"

He felt hungry so he made some toast and honey. CRUNCH. The toast was too hard and made his tooth hurt.

He made a cup of tea. OUCH! The tea was too hot and made his tooth hurt even more.

"I know, I'll have some ice cream. That should make me feel better."

Neddy took a big scoop of honey and peanut ice cream out of the freezer.

OUCH! OUCH! OUCH!

The ice cream was cold and now the pain was terrible. Neddy felt really miserable.

"I shall go out to see my friends to take my mind off the pain," he decided.

He put on his coat and hat and wrapped his scarf round his face.

The ducks were getting their breakfast of pond weed from the bottom of the pond.

"Quack, Neddy, how are you?" they asked.

"Mumble, mumble, mumble," mumbled Neddy, his scarf in his mouth.

"Quack, what? What did you say?" asked the ducks.

"Do you have trouble with your teeth?" asked Neddy.

"We have beaks to help us eat. We don't have teeth," replied the ducks.

A frog sitting on a stone stuck out his tongue and trapped a fly on the end of it.

"I expect frogs don't get tummy ache either when they swallow their food whole," Neddy thought sadly.

Neddy walked on down the lane to the farm. The chickens were having their breakfast. They were pecking at the grains of corn on the ground.

"Do you have trouble with your teeth?" asked Neddy.

"We have beaks to help us eat. We don't have teeth," replied the chickens.

Neddy walked by the field where the cows were lying down, chewing silently.

"Do you have trouble with your teeth?" Neddy asked the cows.

The cows were too busy chewing the grass for a second time to talk to Neddy. Their big jaws moved from side to side.

"No-one seems to be able to tell me about teeth," thought Neddy. "I might as well go home."

On his way home Neddy saw Bonny Bunny.

"Do you have trouble with your teeth?" asked Neddy.

"I have to gnaw at crisp vegetables to keep my teeth sharp like chisels," said Bonny showing Neddy her large front teeth.

Then Neddy saw Clarrisa Cat.

"Do you have trouble with your teeth?" Neddy asked.

"My teeth are purr-fect," Clarissa purred. "They are clean and white and pointed."

"Perhaps you can tell me who can help me with my teeth," said Neddy. "My tooth hurts."

So, Clarrisa took Neddy to Mr Crocodile the dentist who told Neddy to 'open wide'.

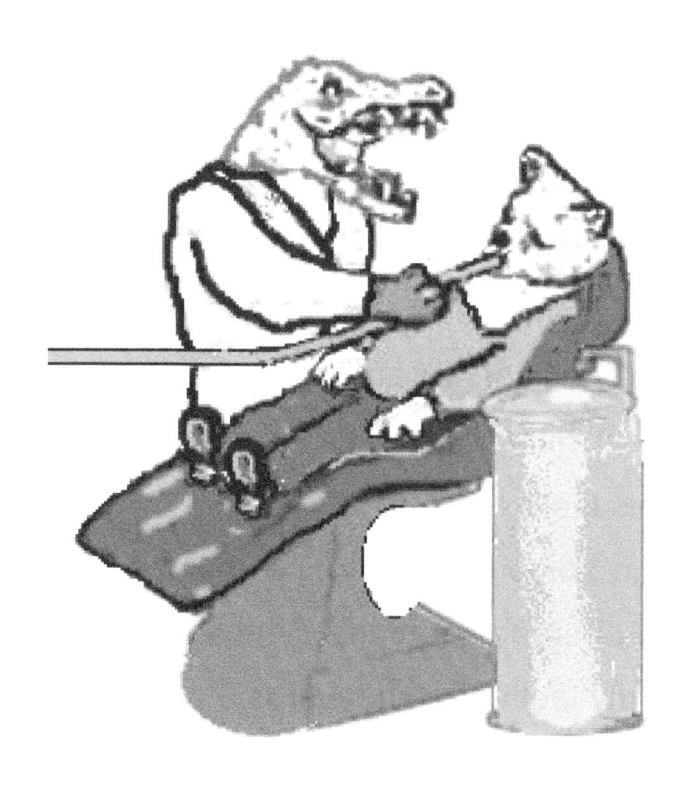

"Dear, dear, dear," said Mr Crocodile. "When did you last clean your teeth?"

Neddy had forgotten to clean his teeth after every meal. What a forgetful teddy.

"You must brush your teeth or eat crispy foods after a meal, drink milk to help make your teeth strong AND visit the dentist every six months," said Mr Crocodile. "Today I shall have to fill the hole in your tooth. That is why you have toothache."

Neddy felt a lot better after visiting the dentist. His tooth did not hurt anymore. Clarissa gave Neddy a blue toothbrush to help him look after his teeth.

"I'll go home now for a nice cup of tea and a slice of bread and honey," he said, "and I'll brush my teeth well afterwards."

NEDDY TAKES ELI TO THE OPTICIAN

Neddy Teddy looked out of his window. He was waiting for Eli Elephant to call at his house. Neddy was taking Eli to the optician's because he needed a new pair of glasses. Neddy saw Eli walking down the middle of the road.

"Oh, no!" cried Neddy. "Eli has forgotten to put his glasses on. He will get run over by a lorry. He can't see that he isn't walking on the pavement."

Neddy rushed out of his house to save Eli. Eli turned left to walk into Neddy's front garden. But he couldn't see Neddy's front gate. CRASH! Eli walked straight into Neddy's fence.

"Oh, dear! That's the second time that Eli has broken my fence," muttered Neddy. "He needs to go to Mr Owl the optician for a new pair of glasses."

Neddy quickly closed his front door and the two friends hurried down the path to the gate. Neddy sighed as they walked past the broken fence. But Neddy was a kind bear and did not say anything to Eli.

Mr Owl the optician tested Eli's eyes. Eli was so big Mr Owl had to make his perch taller so that he could see into Eli's eyes. Mr Owl asked Eli to read the letters on the big poster on the wall. He put lenses in front of his eyes so that Eli could see better.

He asked which circles looked clearer. Mr Owl shone a bright light into Eli's eyes so he could see inside his eyes to make sure that everything was alright. Finally, Mr Owl did something strange that made Eli jump with surprise. He nearly knocked Mr Owl off his perch. A machine puffed air at Eli's eyes. The reading told Mr Owl that the insides of his eyes were healthy.

Then Eli had to pick the frame for the lenses. Eli looked in the mirror.

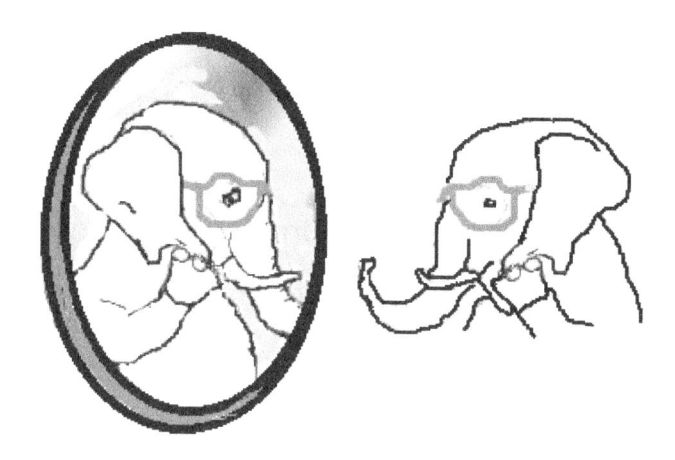

"Have you decided upon the frames you want?" asked Neddy.

"I like these glasses that don't have a frame," Eli said.

"But you won't see them. You need a bright frame so you can see them before you put them on," suggested Neddy.

"You are a good friend," said Eli. "I'm so glad you came with me to help me choose which frames would suit me."

Eli chose a nice bright blue frame for his glasses. He was told to collect them in a fortnight's time.

"I think it's time to go home now," suggested Eli. "But before we go home Neddy, have you made an appointment with Mr Owl to check your eyes?"

What a forgetful teddy! Neddy had forgotten to make an appointment. He was an elderly bear so he should get his eyes checked regularly.

"You are quite right, Eli. I'll make an appointment now," said Neddy. "Then we can go home for a nice cup of tea and a slice of bread and honey."

NEDDY HAS THE FLU

The sun shone brightly through Neddy's bedroom window. Neddy did not feel like looking through it that morning. He staggered out of bed and looked in his mirror.

Neddy stuck out his tongue and it looked fuzzy and pale green. Not clean and pink. It looked awful.

His nose was dripping and yet it felt blocked up. Not clear and letting in the fresh air. He blew his nose into his handkerchief. His handkerchief soon got wet. His nose was red and sore. His eyes were bleary. He could barely keep them open. His throat hurt. He felt hot and sweaty. His whole body ached. And worse of all his head felt that he had a hammer inside it. He had an awful headache.

"What is wrong with me?" Neddy asked himself. "I had better look in my medicine cabinet and see if there's anything in there that will help me feel better."

Neddy staggered to his bathroom and unlocked his medicine cabinet. There were bottles without labels.

"They are not much use," he declared.

There were boxes of medicines that were out of date.

"Better throw those away too. Give them back to the chemist. I feel too ill to go out to buy any medicines. I'll phone Clarissa. She will know what I can do."

"Oh, dear. Poor Neddy," said Clarissa. "I think you must have the flu."

Neddy groaned. Poor Neddy felt so ill all he wanted to do was to go back to bed.

"I'll come and see you," said Clarissa. "I'll go to the chemist's first to get you some medicines."

Clarissa Cat went to the Chemist and bought some medicine to help Neddy feel better. She also bought a thermometer to take his temperature.

She then went to the supermarket and bought some oranges, grapes and lemons.

"Fruit has lots of vitamin C that helps you when you have a cold or have flu," said Clarissa to herself. "I know Neddy will have some honey to sooth his throat."

Neddy was pleased to see Clarissa.

"You are such a good friend," he mumbled under his blankets.

"Let's take your temperature," said Clarissa. "If it is high you have got the flu."

Sure enough, Neddy had the flu.

Clarissa told Neddy to keep warm in bed and rest. She told him to drink plenty of water and take the medicines. Eat the fruit too.

"But how did you catch the flu?" asked Clarissa.

"I got wet and cold waiting for the bus in the rain. I had forgotten to take my umbrella."

What a forgetful teddy!

"Someone in the bus queue was coughing and sneezing," said Neddy.

"That's how you catch germs from other folks," said Clarissa. "That's why you should use a handkerchief when you cough and sneeze so you don't spread the germs."

Clarissa made him a hot drink with lemon juice and some honey.

"There, there," said Clarissa, as she tucked Neddy into bed. "I'll call round to see how you are this evening. Then I'll make you something to eat."

A few days later, with Clarissa's care, Neddy felt so much better. He was able to sit in his armchair and have a nice cup of tea and a slice of bread and honey.

MILLY AND BILLY NEED A VACCINATION

'Ding, dong!'

"That's my front door bell! I wonder who that is?" Neddy said to himself.

He looked out of his window and saw Mrs Mouse. Mrs Mouse lived two doors away.

Neddy would often look after Milly and Billy Mouse when Mrs Mouse went shopping.

"I have to take Milly and Billy to the doctor's this afternoon to get their jabs," said Mrs Mouse.

"Oh!" replied Neddy. "You mean their vaccination? I don't think they will like that very much."

"That's why I've come to ask you to help me," said Mrs Mouse.

"Oh!" replied Neddy again. "In what way?"

"I don't think I can manage both of the twins together. They have been to the doctor's before when they were very little but now they might get upset."

Neddy thought that Milly and Billy were still very little. But he was a kind bear and didn't say anything.

"Would you come with me and hold Billy's hand? I can hold Milly's hand," said Mrs Mouse.

"Of course I'll come along," Neddy said with a smile.

"I'll tell them that we are going to the playground," said Mrs Mouse.

"I don't think that's a good idea," said Neddy. "They will realise that you have told them a lie and then they will never believe what you say."

"What do you think I should say?" asked Mrs Mouse.

"Sometimes it's best to say nothing. Just don't make a fuss," replied Neddy. "We can go to the playground afterwards. I can buy ice creams for all of us," suggested Neddy with a smile. Neddy liked ice cream.

"Well thank you, Neddy. Can you please come and fetch us about two o'clock? That will give us plenty of time to walk to the doctor's in time for the appointment."

Off they all went at two o'clock.

Billy sang, "We're all going to the playground."

"We just need to call in to the doctor's first," said Mrs Mouse. "We won't be long. You need to have a vaccination to make sure that you don't get poorly when you go to school."

Milly looked very worried and started to cry.

"Don't cry," said Neddy. "We are going to the playground afterwards. Then we can all have an ice cream each."

"Hooray," shouted Billy.

They all went into the doctor's for the little mice's vaccination.

Billy and Milly were very brave. It did not take long to get their vaccination.

"Now you won't get poorly with any nasty illnesses," said the nurse.

"No illnesses that can give you spots and make you scratch," said Neddy cheerfully.

"While you are here," said the nurse to Neddy, "would you like to make an appointment to have a vaccination against catching the flu?"

Neddy had forgotten to get a flu vaccination. Which is why he had caught the flu. **What a forgetful teddy!**

When they got to the playground there was an ice cream van. It played a merry tune.

"Ice creams for everyone," said Neddy.

"Thank you, Neddy. You have been very helpful," said Mrs Mouse."

"No trouble at all," replied Neddy. "But I think it's time for me to go home now for a nice cup of tea and a slice of bread and honey."

NEDDY HELPS PERCY TO GET FIT

"Rrrring, rrrring."

"Why is my alarm clock ringing so early?" asked Neddy.

He still felt tired. He rubbed his eyes and yawned.

'Ding, dong!'

"That's my front door bell!"

Neddy got out of bed and looked through his bedroom window. It was Percy Pig.

"What do you want, Percy?" Neddy shouted down from his window.

"You said that you would help me," Percy shouted back. "I'm going to take part in the school's Sports Day. You told me to come round early."

Oh, dear. Neddy had completely forgotten about Percy's school's Sports Day. **What a forgetful teddy!**

"Just a minute and I'll let you in," said Neddy. "I need to get a wash and have some breakfast first."

"You might as well start now," Neddy ordered Percy as Neddy munched on his toasted bread and honey. "Jump up and down for ten jumps, then listen to your heart. Is your heart beating faster or slower?"

Percy jumped up and down. But he started to puff and blow.

"I'm out of breath," he croaked. "My heart is beating really fast. But…"

"Ok, sit down for a bit," suggested Neddy. "Then see if your heart is still beating fast or has slowed down."

"My heart is still beating fast," said Percy.

"Oh, dear!" said Neddy. "You are unfit. We must get you to do more exercises."

"But why…?" Percy started to ask.

"Come along now," interrupted Neddy. "I think we need to see Tina Teddy and borrow a couple of things from her to help you."

So off they went to see Tina Teddy.

"Could we borrow your bicycle and your skipping rope?" asked Neddy.

"Well, yes. I suppose so," replied Tina. "What do you want them for?"

"Percy needs to get fit so he can take part in the school's Sports Day," replied Neddy.

"But, Neddy...,"Percy began.

"Let's go to the park," suggested Neddy, ignoring Percy. "Come along now. It's a lovely day to do some exercises."

First, Neddy got Percy to do some skipping.

"Girls skip," said Percy.

"So do boxers training to fight," replied Neddy. "Skip faster so your heart will beat faster."

Percy started breathing faster.

"I feel hot and I'm getting sweaty."

"You can ride round the park on Tina's bike now," ordered Neddy. "Pedal faster it will make your heart beat faster."

"How are you feeling now?" asked Neddy.

Percy was puffing and blowing.

"I'm tired," answered Percy.

"It's no wonder you get out of breath," said Neddy. "You don't exercise enough."

Percy looked sad. "I don't really know why I need to exercise so I can take part in the Sports Day."

"Well," said Neddy, "I'm afraid you won't be able to win any races that's for sure."

"But I'm not running any races," replied Percy. "I've been trying to tell you I'm just going to write down who the winners are!".

"Oh! Sorry, Percy. How silly of me. I've misunderstood what you were going to do at the school's Sports Day!"

Neddy had forgotten that they need folks to help during Sports Day. **What a forgetful teddy!**

"I'm sorry too," replied Percy. "I didn't explain myself very well."

"Never mind," said Neddy. "Let's go back to my house and have something to eat."

"Yes," agreed Percy, "and we can have a **small** slice of bread and honey and a nice cup of tea, too."